Wloussagnon Raphael Sow
Mambé Auguste-Denise, Boye
Rosin Don-Rodrigue Voko Bi

Agroecology of endogenous mycorrhizal fungi

Wloussagnon Raphael Sow
Mambé Auguste-Denise, Boye
Rosin Don-Rodrigue Voko Bi

Agroecology of endogenous mycorrhizal fungi

cultivation systems of Hevea brasiliensis (rubber tree) (Wild. ex A. Juss, 1865) in Zagné (Western Ivory Coast)

ScienciaScripts

Cover image: www.ingimage.com

This book is a translation from the original published under ISBN 978-620-6-72863-4.

Publisher:
Sciencia Scripts
is a trademark of
Dodo Books Indian Ocean Ltd. and OmniScriptum S.R.L publishing group

120 High Road, East Finchley, London, N2 9ED, United Kingdom
Str. Armeneasca 28/1, office 1, Chisinau MD-2012, Republic of Moldova, Europe
Printed at: see last page
ISBN: 978-620-8-31326-5

Signing sessions

I dedicate this memoir :

To my brothers and sisters,

SOW Alfred, SOW Paterne and SOW Gnonhannouin Blanche, thank you for your daily support.

To my loved ones,

-SOW Kouet Maurice, my loving father. You were a man of values and virtues.

-Tahouo Nowigninon Marie, my mother, thanks to you I was born, your memory will remain indelible in my life. You left us without seeing us through to the end of this effort.

-SOW Fabien, my late great-brother, I copied your courage.

Foreword

The Bioressources et Agronomie master's program at the UFR Agroforesterie, with the option of improving agricultural resources, includes a practical internship to complement the theoretical courses. This internship enables students to familiarize themselves with scientific research. It culminates in a dissertation, which is defended in public. It is within this framework that we have initiated this work, the theme of which is: "Agroecology of endogenous mycorrhizal fungi in *Hevea brasiliensis* (rubber) (Wild. ex A. Juss, 1865) cropping systems in Zagné (western Côte d'Ivoire)".

Thanks

This work was carried out at the Université Jean Lorougnon Guédé (UJLoG). It is with great pleasure that we reserve these lines as a sign of gratitude and recognition to those who have contributed in any way to its elaboration.

My warmest thanks go to Madame TIDOU Abiba Sanogo épouse KONE, Full Professor of Ecotoxicology, President of the Université Jean Lorougnon Guédé, for her dedication and involvement in the training of learners.

I would like to express my sincere thanks to Mr KONE Tidiani, Professor of Hydrobiology, Vice-President of the Université Jean Lorougnon Guédé, in charge of Pedagogy, Research and Technological Innovation, and Mr AKAFFOU Doffou Selastique, Professor of Plant Genetics, Vice-President in charge of Planning, Programming and External Relations, for their commitment to the success of our students.

My sincere thanks go to Mrs TONESSIA Dolou Charlotte, Senior Lecturer in Plant Pathology, Director of the UJLoG Agroforestry Training and Research Unit (UFR), for her teaching, availability and advice.

To Mr. SOUMAHIN Eric Francis, Assistant Professor in Plant Physiology at the UJLoG and head of the Bioressources-Agronomie course, for his availability, rigor and great dedication to the success of the students in this course.

To Mr BOYE Mambe Auguste-Denise, Senior Lecturer in Plant Physiology, for agreeing to act as Scientific Director of this dissertation. His advice, recommendations and great concern have been very useful to me.

To Mr VOKO Bi Rosin Don Rodrigue, Maitre-Assistant en Microbiologie des Sols, for his technical supervision of this work. His advice and recommendations were very useful. I would like to express my gratitude to him once again.

To Mr. BAYALA Jacques, Site Manager of Compagnie Hévéicole du Cavally (CHC), for accepting me into the company he runs.

The manipulation part of this work was carried out at the Laboratoire de Biotechnologies Végétales et Microbiennes (LBVM) of the Institut Nationale Polytechnique Felix Houphouet-Boigny. For this, I would like to thank Mr. ZEZE Adolphe, Professor of Microbial Ecology and Director of the aforementioned laboratory.

I would also like to thank all the teaching staff at UJLoG, in particular those at the Agroforestry UFR, as well as the administrative and technical staff.

A very warm thank you to all INPHB PhD students, in particular SEKA Saustène and AKAFFOU Marie-Ange for their assistance during the microbiological analyses.

Table of contents

INTRODUCTION

Hevea brasiliensis is a plant native to Brazil. Its cultivation began in South-East Asia and now extends throughout the humid tropics (Compagnon, 1986). It was introduced to Côte d'Ivoire in 1956 by the Société Africaine de Plantation d'Hévéa (SAPH) as part of the diversification of export crops (Dusotoit-Coucaud, 2012). Hevea is a tree with high value-added by-products. It is the only commercially exploited source of natural rubber, indispensable in the tire industry and in everyday life. Rubber tree cultivation, which began in 1951, is steadily increasing. According to SIPH (2013), over 70% of natural rubber is used to manufacture tires for the automotive industry. This high proportion of dry rubber in tires and its diverse uses has increased global demand over the last fifteen years (Dusotoit -Coucaud, 2012).

Côte d'Ivoire has been ranked the top rubber-producing country in Africa since 1997 and sixth in the world since 2018 (Kouassi, 2018). In 2019, it produced 664,695 tonnes of natural rubber (FAOSTAT, 2020). Heveaculture is of major economic interest in Côte d'Ivoire. It ranks fourth among export crops after cocoa, coffee and cotton, and generates an annual income of 203 billion CFA francs (Attobra, 2010). Rubber tree cultivation in Côte d'Ivoire requires ever larger areas. Indeed, the area under rubber plantations rose from 116,000 ha in 2002 (MINA, 2002) to over 318,000 ha in 2014 (CNRA, 2015) and then to 534,000 ha in 2018 (Kouassi, 2018). This increase in arable land is accompanied by the massive use of chemical inputs (fertilizers, pesticides and herbicides) with the aim of improving soil fertility and thus increasing production. However, such cultivation practices could have harmful consequences for the environment and compromise sustainable rubber cultivation in Côte d'Ivoire. As a result, this crop is limited by a number of factors, including soil and climate hazards, disease, physical degradation and declining soil fertility (Piéri, 1998).

Alternative solutions for sustainably improving rubber tree production could involve harnessing the beneficial effect of arbuscular mycorrhizal fungi (AMF). These soil fungi form symbiotic associations with the roots of most terrestrial plants, forming mixed organs known as mycorrhizae. These mycorrhizae improve water and mineral uptake, and plant resistance to biotic and abiotic stresses (Smith & Read, 1997).

Today, several CMA inocula have been developed worldwide to ensure sustainable cultivation of a wide range of agricultural crops. Integrating this technology into agriculture in general, and Ivorian rubber cultivation in particular, requires the design of inocula based on endogenous strains adapted to local ecological conditions. Implementing this innovation requires prior knowledge of the life cycle of AMCs and their ability to colonize and then benefit plants. The ability of CMAs to colonize and benefit plants depends on the physico-chemical properties of soils and the cultivation techniques applied, such as the use of chemical inputs and ploughing, which have a negative influence on CMA development (Duponnois *et al.,* 2011; Voko *et al,* 2013).

Thus, before considering the introduction of mycorrhization technology in Côte d'Ivoire rubber cultivation, it is necessary to know the behavior of endogenous CMAs in the context of rubber cultivation in Côte d'Ivoire. The general objective of this study is to determine the potential of endogenous CMAs in soils under rubber cultivation.

The specific objectives are as follows:

-Determine the physicochemical characteristics of soils under rubber cultivation;

-determine the abundance and activity of endogenous CMAs;

-highlight the links between plant health and CMA activity.

The first part of this thesis is devoted to general information on arbuscular mycorrhizal fungi and rubber cultivation. The second part presents the equipment and methods used to carry out the work in the laboratory and in the experimental plot. The third part presents the results obtained, followed by a discussion. Finally, a conclusion and perspectives will constitute the last part of our manuscript.

Part 1: GENERAL

1. Ecology of mycorrhizal fungi

1.1. Different types of mycorrhizae

The word mycorrhize is of Greek origin. It expresses the collaboration between a fungus (myco) and the roots (rhizes) of a plant. This collaboration is in fact a symbiosis, as it results from a mutual agreement between the two organisms. It's based on the fact that both partners derive benefits from the relationship. The fungus draws sugars from the plant, while the plant receives minerals and water from the fungus. Indeed, without this association, the mycorrhizal fungus cannot complete its life cycle Lambers *et al.* (2009). The classification of mycorrhizae is based on the type of fungus associated with it, depending on whether it is aseptate, i.e. a zygomycete of the order Glomales, or septate, like the ascomycetes or basidiomycetes (Smith & Read, 2008).

Ectomycorrhizae account for around 5% of terrestrial species and some 5,000 species of fungi. They are nutrient exchange sites, mainly for nitrogen and phosphorus. This symbiosis makes it possible to reduce pollution and the use of fertilizers and phytosanitary products. They occur mainly in trees and shrubs (Egerton-Warburton & Allen, 2001) and involve a wide variety of Basidiomycetes, Ascomycetes and Zygomycetes fungi. During the interaction, the fungal partner forms a sleeve of hyphae around certain roots, and insinuates itself between the cells of the epidermis and outer cortex without ever penetrating the plant cells (Smith & Read, 2008).

Ectoendomycorrhizae are similar to ectomycorrhizae in the presence of a large external superficial mycelium, but also possess certain intracellular formations. The fungus forms a veritable fungal sleeve, but its Hartig network is poorly developed. From this network, hyphae penetrate the interior of living cortical cells, where the fungus develops platoons endowed with cytoplasmic organelles Dos Santos *et al.* (2001).

There are two types of endomycorrhizae: platy endomycorrhizae, which are unhealthy, and arbuscular endomycorrhizae, which account for 95% Redecker *et al.* (2000). The

Arbuscular endomycorrhizae or arbuscular mycorrhizal fungi (AMF) have many benefits for the plant and are dominant on young seedlings (Egerton-Warbuton & Allen, 2001). The use of AMCs enables a high nutrient supply, thanks to the large surface area covered by the hyphae, which are very small, long and abundant, unlike absorbing hairs. In endomycorrhizae, the fungus's subterranean mycelial filaments (hyphae) do not form a sleeve around the root, but pass through the walls of the root cells, hence the name "endomycorrhizae", reflecting the presence of the fungus inside the host plant's cells. Arbuscular mycorrhizal symbiosis (AM) is probably the most widespread terrestrial symbiosis. In fact, it is associated with more than 80% of today's terrestrial plant species, mainly herbaceous species, including important cultivated species (wheat, rice and soybean) (Smith & Read, 2008). (Selosse & Le Tacon, 1998) have highlighted different types of mycorrhizal associations (see Figure 1).

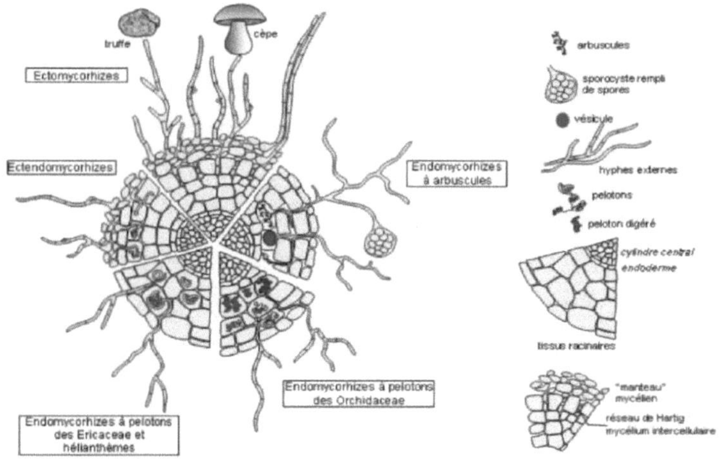

Figure 1: Diagram of the different types of mycorrhizal associations (Selosse & Le Tacon, 1998)

1.2. Morphology and reproduction of arbuscular mycorrhizal fungi (AMF)

MACs are multinucleate eukaryotes. They are formed of cenocytic hyphae that lack septa, allowing the nuclei to move freely within the mycelium SchdBler *et al.* (2001). Their morphology has changed little since they first appeared, since the arbuscules found in fossils dating back over 400 million years resemble those observed in living organisms today. There is great diversity in the structure and shape of CMA spores (Figure 2). Spores are produced by CMAs to ensure their progeny through asexual reproduction. Sexual reproduction in CMA has not yet been proven (Redecker & Raab, 2006).

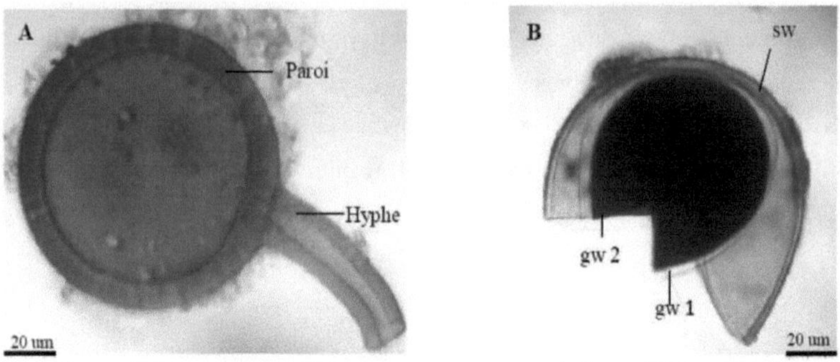

Figure 2: Examples of CMA spore structure and shape (Voko, 2014).

A: Glomus sp. isolated spore, glomoid form

B: Isolated Acaulospora colombiana spore, acaulosporoid form,

1.3. Classification of arbuscular mycorrhizal fungi

The taxonomy of CMAs was initially based on spore morphotyping, taking into account both structural similarities between spores and phylogenetically important characters (Morton & Benny, 1990). Recent morphological and phylogenetic studies have made it possible to group all arbuscular mycorrhizal species into a new phylum, the Glomeromycota, divided into 4 orders comprising 10 families and 15 genera and around 200 species. All of these are associated with some 225,000 terrestrial plant species. The following phylogenetic tree shows the higher taxa within the class Glomeromycetes (Table I).

Table I: Classification of arbuscular mycorrhizal fungi (AMF) according to Redecker *et al.* (2013)

Phylum	Classe	Ordre	Famille	Genre
Glomeromycota	Glomeromycètes	Diversisporales	Diversisporaceae	Tricispora Otospora Diversispora Corymbiglomus Redeckera
			Acaulosporaceae	Acaulospora
			Sacculosporaceae	Sacculospora
			Pascisporaceae	Pascispora
			Gigasporaceae	Scutellospora Gigaspora Intraornatospora Paradentiscutata Dentiscutata Cetraspora Racocetra
		Glomerales	Claroïdeoglomeraceae	Claroïdeoglomu
			Glomeraceae	Glomus Funneliformis Septoglomus Rhizophagus Sclerocystis
	Archaeosporomycètes	Archaeosporales	Ambisporaceae	Ambispora
			Geosiphonaceae	Geosiphon
			Archeosporaceae	Archeospora
	Paraglomeromycètes	Paraglomerales	Paraglomeraceae	Paraglomus

1.4. Life cycle of arbuscular mycorrhizal fungi

The mycelium of arbuscular mushrooms is cenotic, i.e. without partitions or cell-forming separations. Along this cenotic mycelial network, the fungus forms spores designed to propagate and disseminate the species. From the extraradicular mycelium, the spores develop germ tubes that can extend several centimetres in the direction of the active roots, resulting in primary root infection. In some species, intraradicular vesicles differentiate in the root cortex and possess properties analogous to those of spores. Dead or living root segments can be a source of inoculum for newly developed roots Duponnois, *et al.* (2011).

1.5. The role of arbuscular mycorrhizal fungi

Arbuscular mycorrhizal fungi play an important role in plant growth and protection, and in improving soil quality. While the results of mycorrhization vary according to the "plant-arbuscular mycorrhizal fungi" association, they almost always translate into a marked increase in plant growth. Mycorrhized plants assimilate phosphorus and sometimes nitrogen more easily, are more resistant to water stress, tolerate limestone better and are less attacked by root parasites (Davet, 1996). Arbuscular mycorrhizal fungi improve plant nutrition by increasing the volume of soil accessible to the plant, and by mobilizing forms of nutrients that the plant would not normally have been able to absorb (Smith & Read, 2008).

Several ecological roles for mycorrhizal fungi have been highlighted. These include the mobilization of nitrogen (N) and phosphorus (P) by organic polymers, and the release of nutrients from mineral particles or rock surfaces by degradation (Finlay, 2008; Smith and Read, 2008). The binding of microaggregates is favoured by the production of large quantities of polysaccharides by the extraracinar mycelium. It has been identified as a homologue of heat shock prote in, a stress protein (Gadkar & Rillig, 2006). CMAs also enable the plant to resist certain bacteria such as Xanthomonas campesris Liu *et al.* (2007). Some authors have also reported that arbuscular mycorrhizal fungi are able to limit nematode damage and thus improve production in banana (Elsen *et al.,* 2008), grapevine (Li et al., 2006), yam (Tchabi, 2008), tomato (Vos *et al.*, 2012) and so on.

Improved plant resistance to bacteria has also been demonstrated. As far as plant resistance to nematodes is concerned, various authors

1.6. Agroecology of arbuscular mycorrhizal fungi

Arbuscular mycorrhizal fungi are obligate symbionts, meaning they are unable to synthesize their own sugars (Brigitta & Sumalan, 2011). They cannot be cultivated in artificial environments and are unable to complete their entire development cycle without forming symbioses with an autotrophic plant. Root colonization of the host plant and the development of arbuscular mycorrhizal fungi are carried out in four stages: (i) the development of colonizing propagules (spore germination),

(ii) the development of structures within the roots playing an important role in morphological characterization, (iii) the development of extraracinar mycelia in the soil and (iv) spore production (Smith & Read, 2008).

1.6.1. Relationships between soil properties and arbuscular mycorrhizal fungi communities

Soil is the habitat of arbuscular mycorrhizal fungi. Physical parameters, organic matter and chemical constituents define soil type and fertility. These parameters directly or indirectly influence their development and their relationship with plants. Soil physico-chemical properties are also influenced by the intensity of land use (Angui *et al.,* 2010; Yao *et al.,* 2010). The diversity and composition of arbuscular mycorrhizal fungi communities are highly dependent on the type and intensity of soil use Oehl *et al.* (2010). Also, increasing organic matter can stimulate saprophytic organisms that parasitize arbuscular mycorrhizal fungi (Lozupone & Klein, 1994).

1.6.2. Impact of cultivation practices on arbuscular mycorrhizal fungi communities

Community composition changes with agro-ecological zones. Arbuscular mycorrhizal fungi diversity varies with succession, crop rotation and agroecological zones Dare *et al.* (2012). Arbuscular mycorrhizal fungi communities are affected by human disturbance, particularly in semi-arid regions Muchane *et al.* (2012). Arbuscular mycorrhizal fungi communities can also vary according to latitude and rainfall regime (Friberg, 2001), as well as over time (Lovelock & Ewel, 2005).

2. General information on rubber trees

2.1 Origin and systematics

The rubber tree is native to Amazonia (South America). It was first introduced to Ceylon in 1876, then successfully planted in the Far East and Africa, according to the Centre Technique Forestier Tropical and the Institut de Recherche sur le Caoutchouc (CTFT & IRCA). Rubber tree cultivation in Côte d'Ivoire began in

1951. Hevea systematics are as follows:

Règne : Végétal
Sous règne : Tracheobionta
Embranchement : Magnoliophyta
Sous embranchement : Angiospermes
Classe : Dicotylédones
Sous classe : Rosidae
Ordre : Euphorbiales
Famille : Euphorbiaceae
Genre : Hevea
Espèce binomiale : *Hevea brasiliensis* (Willd. ex A. Juss) Müll. Arg

2.2. Botanical characteristics and morphological description of the rubber tree

The Hevea is a plant with a trunk that bulges slightly at the base, and three-leaf leaves supported by a long petiole (Figure 3). The tree's flowers are unisexual and produce fruits called capsules. These capsules have three chambers, each containing a seed. The bark, which is smooth and around 1 cm thick, contains the laticifere vessels (specialized cells in which latex is synthesized) (Panthou, 2013). Latex is composed mainly of rubber particles (C5⅛) but also of resins, proteins and enzymes and sometimes even starch. The underground system consists of a pivot and lateral roots. Vegetative propagation is by grafting.

In its natural Amazonian environment, the rubber tree is a large forest tree, commonly exceeding 25-30 m in height and measuring over 1 m in circumference. The largest trees can be around fifty meters (Obouayeba, 2006). The trunk is generally straight (conical at the base for Seedlings and cylindrical for grafted plants). The wood is homogeneous, soft and brittle, with a sensitivity to wind damage that varies from wind to wind and from clone to clone (Demol, 2002). The bark is grey and represents the capital of rubber cultivation. It is the most important part for the planter, as it is the most exploited. It is made up of two parts: the hard outer bark and the soft inner bark. When mature, it dries, then turns yellow and bursts along the median line of each compartment. The flowers are small, light-yellow and clustered (Demol, 2002). Leaves are formed in stages at the top of each growth unit. The fruit is a generally three-lobed capsule, each containing one seed. The fruit shell reaches maturity three months after fertilization.

2.3. Economic importance

The rubber tree is cultivated for its latex, which is used to manufacture natural rubber. At the end of the exploitation period, the tree can also be used as timber and firewood Obouayeba *et al.* (2006). Flexible and resistant to abrasion, natural rubber is an indispensable material for the manufacture of tires, hoses, conveyor belts and pump equipment. Natural rubber's resistance to water and most liquid chemicals makes it exceptionally well suited to diving equipment, chemical piping and more. Thanks to its elasticity, natural rubber is an important material for various shock-absorbing and vibration-reducing systems. In addition to rubber, rubber trees can also supply secondary products. For example, rubberwood can be used as firewood or timber. Seed cake can also be used as animal feed.

Figure 3: Hevea brasiliensis (rubber tree) plantation: Source: https://thinglishlifestyle.com/.

2.4. Ecological requirements

Hevea requires a warm, very humid climate. It grows best in deep, loose, well-drained soils. It prefers sandy-clay or sandy-clay soils. Due to its climatic requirements, rubber cultivation is practiced in Latin America, Southeast Asia, West and Central Africa. Hevea is a heliophilous plant and cannot withstand strong winds. For good growth and production, rubber trees need an average annual temperature of between 25 and

37°C, a minimum annual sunshine duration of 1,650 hours, and annual rainfall of between 1,500 and 2,500 mm. The soil must be loose, with a pH of between 4.5 and 6, sandy-clay to clayey in texture, deep and not waterlogged to a depth of at least one meter. Cultivation requires flat or gently sloping ground (less than 8%) to avoid erosion, but in the case of steeply sloping ground (greater than 8%), terraces can be built on a contour Obouayeba *et al.* (2006).

2.5. Pests and diseases

Fungi cause various diseases in Hevea brasiliensis. Among these diseases, white rot caused by Rigidoporus lignosus or Fomes lignosus causes enormous damage to rubber plants worldwide, and particularly in Côte d'Ivoire. This disease can cause over 60% mortality of rubber trees in a plantation in the absence of control measures (Wahounou, 2010). In addition to this disease, others are caused by a variety of fungi. Among these, Heminthosporium hevea, frequently found on young nursery plants, causes rounded yellow spots on the leaves. The Phytophthora palmivora fungus causes burns on petioles, leading to early leaf drop. As for Colletotrichum gloesporioides, it causes very severe attacks on rubber trees, resulting in the fall of affected young leaves (Ndong, 2007). The fungus Rhizoctonia solani causes the appearance of spots of varying size and shape, the appearance of which depends essentially on the stage of development they have reached.

Part 2: MATERIAL AND METHODS

1. Study site and plot characteristics

1.1. Location of study site

The study was carried out at the Compagnie Hévéicole du Cavally (CHC) site. This agro-industrial enterprise is located 160km from Daloa and 573km from Abidjan, in the Department of Taï, more precisely in the sub-prefecture of Zagné between 6°13'N latitude North and 7°29'W longitude West. The area's average temperature varies between 18 and 36°C. The climate is hot and humid, with rainfall ranging from 1,000 to 1,500 mm/year. Two rainy seasons (April-July; September-November) and two dry seasons (December-March; July-September) give it a homogeneous vegetation with a dense, humid forest containing a diversity of forest species. The area is also conducive to the development of several crops, including coffee, cocoa and, more recently, rubber, oil palm, food and vegetable crops (Neobot, 2020).

1.2. Plot characteristics

Compagnie Hévéicole de Cavally (CHC) is an agro-industrial enterprise comprising a factory and a rubber plantation covering six thousand hectares divided into three main plantation sections: Dazahon (06°54.564'N, 006°26.263'W), Bohié (06°08.705'N, 007°38.808'W) and Doni (06°07.763'N, 007°43.068'W). Each of these sections corresponds to a planting area appropriate to the rubber variety. The planting density is 510 trees per hectare, with a technical itinerary of 2.8 metres between rows and 7 metres between rows. The company has a nursery plot and graft wood consisting of several clones. The planting plots differ in the type of clone present in each of them. The study was carried out on plots in the three sections according to planting age. Figure 4 shows a map of the sampling area.

According to the company's data, YARAMILA fertilizer (NPK formulation 23/10/05) is systematically applied to each rubber tree from the 2ème month until the 6ème year of cultivation. The total amount of fertilizer received by each tree during the six (6) years of cultivation is 9000 g/tree (Table II).

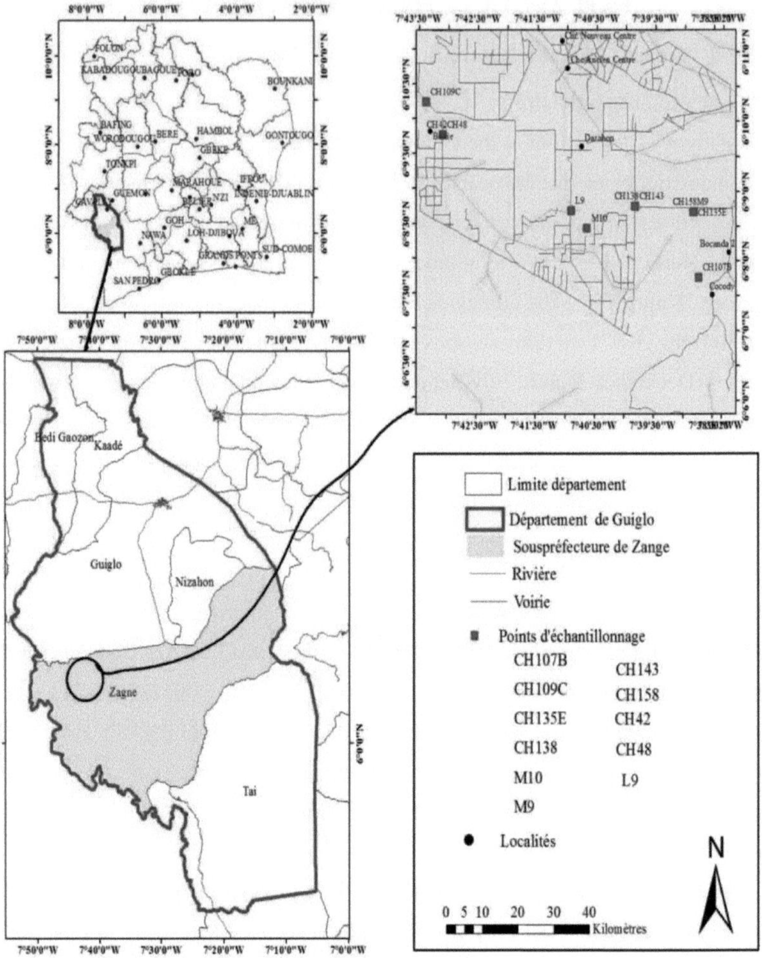

Figure 4: Map of the Département de zagné showing the study area and sampling points

Table II: Fertilizer use on Compagnie Hévéicole du Cavally plots

Age of Parcel (years)	Type of fertilizer	Formulation (NPK)	Application technique	Quantity of fertilizer (g)	
				First tower	Second tower
1	Yaramila	23/10/05	In the round	0	200
2	yaramila	23/10/05	In the round	250	250
3	yaramila	23/10/05	In the round	250	250
4	yaramila	23/10/05	In the line	300	300
5	yaramila	23/10/05	In the line	300	300
6	yaramila	23/10/05	interlining	300	300
	yaramila	23/10/05	interlining	300	300

2. Study materials

2.1. Plant material

The plant material consisted of two Hevea brasiliensis clones, IRCA41 and PB217 from different growing years: 2017; 2015; 2009 and 1998. Infected roots and organ parts of these plants were collected for characterization of arbuscular mycorrhizal fungi (AMF), followed by characterization of infection-associated microorganisms.

2.2. Fungal material

Fungi isolated from the soil beneath rubber plants.

Figure 5: Soil samples taken from under a rubber tree (clone PB217)

2.3. Technical equipment

An autoclave and then an oven were used to sterilize the substrate. A series of four sieves with mesh sizes of 500 µm, 125 µm, 90 µm and 45µm, and beakers were used for spore extraction by wet sieving. Estimation of mycorrhization intensities required the use of hot plates and chemicals such as trypan blue, potassium hydroxide and hydrochloric acid. Standard microbiology laboratory equipment was also used to analyze infected organ parts. This equipment consisted of glassware (Petri dishes, test tubes, hemolysis tubes, 2 mL Eppendorf tubes, beakers, 10 and 100 mL graduated test tubes), an OHAUS Explorer electronic balance accurate to 0.01 g for weighing, a water bath for solubilizing culture media, an autoclave for sterilizing media and equipment, a hood and Bensen nozzle for aseptic working conditions, an incubator for incubating Petri dishes and a Motic optical microscope for microscopic visualization.

3. Study methods

3.1. Plot layout

The study was carried out on a total of 12 single-celled plots ranging from 3 to 15 ha in size, of two rubber varieties, spatially distributed as shown in Figure 4: plots L12, M10, CH107B, CH109C, CH138 and CH42 of the IRCA41 variety, and L9, M9, CH135E, CH158, CH143 and CH48 of the PB217 variety. The plots are divided into 3 age groups:

- Under 3 years for L12, M10, L9, M9,
- Between 3 and 5 years for CH107B, CH109C, CH135E and CH158

- Over 5 years for CH138, CH42, CH143 and CH48

2.1. Determination of the physico-chemical properties of soils under rubber cultivation

2.1.1. Soil sampling

In each plot, soil samples were taken at four separate points (four incremental samples per plot) at least 100 m apart, using the method described by (Haung & Cares, 2004). Soil from each excavation (500 g per excavation) was collected in a bucket and mixed to form a composite sample (2 kg) of the four incremental samples. This mixture (composite sample) was transposed into a polyethylene bag coded according to the plot. The material was rinsed with water and then disinfected with 10% bleach at 12° before being transferred from one sample to another. The soil was taken from the foot of the rubber plants at a depth of 16 cm using a daba. It has been shown that MACs colonize only the young, fine roots, so 95% of the absorbent root hairs are concentrated in this horizon (Burle, 1961). For each soil sample, 100 g were taken and sent in the raw state to the Laboratoire de Pédologie for soil analysis at the Institut Nationale Polytechnique Houphouët-Boigny.

2.1.1.1. Particle size measurement

Particle size analysis was carried out using the densimetric method with a Robinson pipette. This method consists in separating soil particles according to their size. It provides a breakdown of fine particles less than 2 mm in diameter into different size classes. The triangle of mineral textures was used to determine the soil texture of the study plots (figure 6).

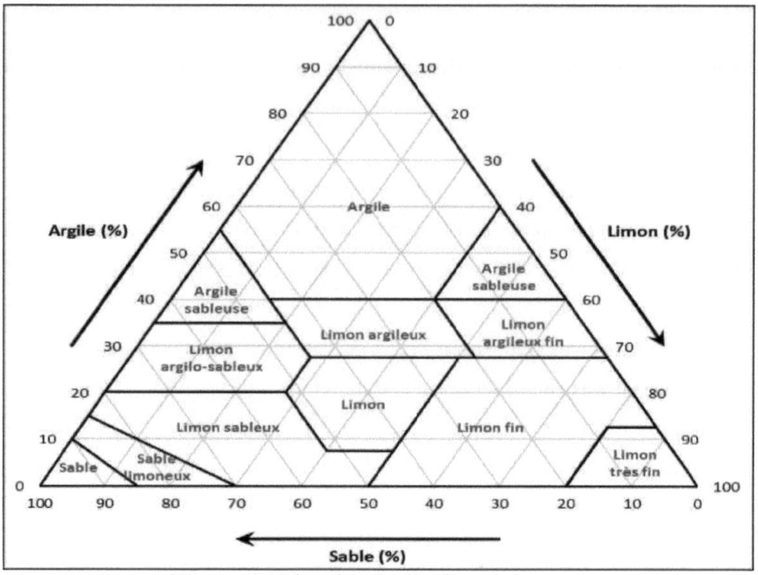

Figure 6: Triangle of soil mineral textures

2.1.1.2. Determining pH

Water pH was measured electrometrically using a pH meter, on a suspension of 20 g of soil in 50 ml of distilled water, after shaking the suspension at 175 rpm for 30 minutes. Hydrogen ions or hydrons (H^+) contained in water. A distinction is made between :

*PH<5: very acidic soil

*PH< 6.5: acidic soil

*PH= to 7 (between 6.5 and 7.5): neutral soil, neither acidic nor alkaline

*PH> to 7.5: alkaline (or basic) soil

*PH> to 8: very basic soil

2.1.1.3. Determining the proportion of organic matter

Organic matter analyses took into account total organic carbon and total nitrogen. Total organic carbon was measured wet at a reaction temperature of 120°C (Walkley & Black, 1934). Total nitrogen was determined by the Kjeldahl method (Bremner,

1960), which involved hot attacking living matter with concentrated sulfuric acid (H_2SO_4) in the presence of catalysts such as potassium sulfate (K_2SO_4), copper sulfate ($CuSO_4$) and selenium (Se); the extraction solution was then distilled with excess sodium hydroxide and titrated with H_2SO_4 (1 N). The standard for interpreting organic matter is: very low OM soil (OM between 1 and 5%), low OM soil (OM less than 5%), medium OM soil (between 5 and 6%), very high OM soil (greater than 7%).

2.1.1.4. Determination of total and assimilable phosphorus content (Pass)

Phosphorus (P), essentially in the form of calcium-bound orthophosphates (apatites), is a relatively scarce major element, with estimated average contents of around 0.1%. Total P and assimilable P are generally determined (Pansu & Gautheyrou, 2003). Total P, comprising organic and inorganic phosphorus, is determined after total wet mineralization by attacking 5 g of ground soil with a reagent composed of 60% perchloric acid, nitric acid (d = 1.4) and distilled water. The assimilable P corresponding to the most available biological P (labile organic and inorganic forms and microbial P) is extracted in sodium bicarbonate at pH 8.5 (Olsen method, 1952).

2.1.1.5. Determination of exchangeable base content

Exchangeable bases (Ca^{2+}, Mg^{2+} , K^+ , Na^+) and CEC ($cmol.kg^{-1}$) were measured in 5 g of soil sample using barium chloride solution as extraction reagent according to ISO 11260 (2018). Cation exchange capacity is determined as the degree or ability of a substrate to store nutrients The standard for interpreting cation exchange capacity (CEC) is : very low CEC soils (CEC below 5 cmol/kg), low CEC soils (CEC between 5 and 10 cmol/kg), medium CEC soils (CEC between 10 and 15 cmol/kg), high CEC soils (CEC between 15 and 20 cmol/kg) and very high CEC soils (CEC above 20 cmol/kg) (Calvet & Villemin, 1986).

3.3 Assessing the health of rubber seedlings

3.3.1. Epidemiological survey and sampling of diseased plants

The study consisted in counting the number of diseased plants in each plot. On each study plot, a sub-plot of 1 ha (510 plants) was delimited. The number of diseased plants per sub-plot was then determined by observing their symptoms.

During the inventory of diseased plants, all accessible parts showing symptoms were sampled and collected in sterile Eppendorf tubes coded according to plot and symptom type.

3.3.2. Microbial diversity of infected organs

3.3.2.1. Preparation of stock solutions and dilutions

The harvested rubber tree organs were cut in the fume hood using a sterile blade into 10 fragments of 2 cm length per sample, then ground in a sterile mortar. Using a 100 µl micropipette, the mother suspension was taken and transferred to an Eppendorf tube containing 900 µl of sterile distilled water to obtain the first dilution, then 100 µl of the first dilution was transferred to a tube containing 900 µl of sterile distilled water to obtain the second dilution and so on up to the fifth dilution.

3.3.2.2. Inoculation of observed microorganisms

The dilutions obtained were inoculated onto agar media specific to the germs of interest. These were aerobic mesophilic germs (AMG), Bacillus, microscopic fungi and Enterobacteriaceae. For GAM, inoculation was carried out on Plat Count Agar (PCA) by incorporation (double layer). Incubation was carried out at 37°C for 24 to 48 hours. Enterobacteriaceae were detected on a double layer of VRBG agar (Violet Red Bile Glucose Agar). Incubation was carried out at 37°C for 24 to 48 hours. Bacillus were counted on Mossel agar incubated at 30°C for 24 to 48 hours. For fungi, 100µl of solution was surface-seeded on Sabouraud chloramphenicol medium and incubated at 37°C for 72 hours.

3.3.2.3. Colony count of observed microorganisms

The number of colonies of microorganisms was determined by simple macroscopic counting on the various Petri dishes. The number of colony-forming units was determined using the following formula:

$$UFC=\frac{\sum C}{v(n_1+0,1n_1)d}$$

With :

C: number of colonies observed on all selected and usable plates (plates from two successive dilutions, at least one of which contains 15 colonies; only plates corresponding to a number of CFU less than or equal to 300 are considered in the calculation) v: volume of the suspension spread on the surface of the media in mL (e.g. 0.1 mL). ni: number of plates retained for the first dilution (the lowest). n2: number of plates retained for the second dilution (the highest).

d: dilution ratio corresponding to the lowest dilution used (d = 1 for the undiluted sample; d = 0.01 for the 1/100 dilution, etc.).

Results are rounded to two significant figures after the decimal point and expressed as CFU/ml in the original cell suspension.

3.4. Evaluation of mycorrhization

3.4.1. Sampling rubber tree roots

The young, fine roots of the seedlings are gently removed with a chisel, minimizing damage to the main root. The chisel blade was rinsed with water and then disinfected with 12° bleach before being passed from plant to plant. These root samples were transposed into polyethylene bags coded according to plot (variety, age, repetition) Voko *et al.* (2013).

3.4.2. Extraction and enumeration of mycorrhizal fungi spores

Extraction of CMA spores from sampled roots was carried out using the wet sieving method described by Gerdemann & Nicholson (1963), followed by the modified sucrose gradient spore separation method of Walker *et al.* (1982). In a beaker, 50 g of soil was suspended in 0.5 L of running water and stirred vigorously for 1 min with a spatula to separate fungal propagules and soil particles. After settling for 30 seconds, the supernatant is passed through a series of four sieves of decreasing mesh size (500,

125, 90 and 45 µm) arranged in this order. This operation is repeated three times for the same soil sample. The contents of the 500 µm mesh sieve are discarded, as they rarely contain spores. Each of the contents of the other three sieves is collected with a little running water using a pissette, then transferred to centrifuge tubes. An initial centrifugation was performed at 2000 rpm for 5 min. At the end of this first centrifugation, the supernatant was discarded and the pellet suspended in 10 ml of 50% sucrose. After shaking, a second centrifugation was performed at 2000 rpm for 1 min. The supernatant is passed through a 45 µm mesh sieve, while the pellet is discarded. The spores are recovered and suspended in approximately 10 ml of distilled water, in three 50-ml beakers.

The enumeration procedure used is that recommended by the INVAM collection, (2018). It is used to determine the abundance (or density) of spores. The contents of each beaker containing the spore suspension are transferred to petri dishes and observed under a binocular loupe at 40x magnification. Spores are first counted in at least 10 fields of view. If there is an average of less than four spores per field (low spore density), an exhaustive count is made of the entire contents of the beaker. Otherwise, the spore count is carried out in 40 fields of view and multiplied by 289 (total number of fields of view in the petri dish). The spore density (spores.g-1) is then determined for each soil sample. It is defined as the number of spores in 1g of dry soil.

$$\text{Densité (spores.g-1)} = \frac{\text{Nombre de spores}}{\text{Quantité de sol}}$$

3.4.3 Evaluation of mycorrhization frequency and intensity

The presence of arbuscular mycorrhizal fungi in the roots is not visible without prior staining. Recovered fine rubber roots were washed with water and stained with a trypan blue solution using the method of Phillips & Hayrnann (1970). The fine roots were cut into 1 cm-long fragments. Ten fragments were heated in a 10% KOH solution at 80°C for one hour, then rinsed thoroughly under running water. Treatment with

KOH removes the cytoplasm from the cells. As a result, vesicles and arbuscules stain better with trypan blue. These fragments were again heated to 80°C in 0.05% trypan blue solution. After staining, root fragments were mounted between slide and coverslip in 50% glycerol to estimate root colonization (Kormanik & McGraw, 1982). Mycorrhization frequency and intensity were assessed under the light microscope using the method of Trouvelot *et al.* (1986), (figure 7).

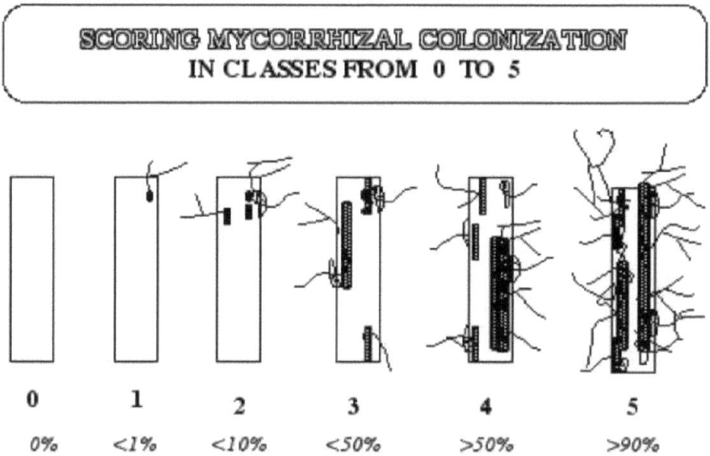

Figure 7: Classification of different levels of mycorrhizal colonization of the root cortex from Trouvelot *et al.* (1986).

2.5. Statistical data analysis

The data collected were subjected to a classification criterion analysis of variance (ANOVA) using SPSS11.5 software at the 5% threshold. A one-way ANOVA was performed to compare means for mycorrhizal propagule abundance, mycorrhizal colonization, production performance, morbidity, mortality and microbial load parameters. In the case of significant differences at the 5% threshold, homogeneous groups were determined using Duncan's method. Principal component analysis was performed to relate physicochemical properties (OM, PH, CEC, C, N and exchangeable bases), mycorrhizal intensity and plant health using R software.

Part 3: RESULTS AND DISCUSSION

1. Results

1.1. Physicochemical properties of study soils

1.1.1. Particle size

Granulometric analysis was used to determine the texture of the soils under rubber cultivation on the CHC plots. The proportions of sand are between 50 and 51%, silt between 13 and 15% and clay between 35 and 36%. In the triangle of soil mineral textures, these proportions correspond to the sandy-clay texture, which reveals that all these plots have a sandy-clay texture (Table III).

Table III: Physical properties of soils under rubber cultivation at Compagnie Hévéicole du Cavally

Plots	Sand (%)	Silt (%)	Clay (%)	Texture
L12	50	15	35	Sandy-clay
M10	50	15	35	Sandy-clay
CH107B	51	13	36	Sandy-clay
CH109C	51	13	36	Sandy-clay
CH138	50	15	35	Sandy-clay
CH42	50	14	36	Sandy-clay
L9	50	15	35	Sandy-clay
M9	50	15	35	Sandy-clay
CH135E	50	14	36	Sandy-clay
CH158	51	13	36	Sandy-clay
CH143	50	15	35	Sandy-clay

CH48	50	14	36	Sandy-clay

1.1.2. Analysis of soil chemical quality

Soil water pH for all plots ranged from 6.1 to 6.7 (Table IV). The soils of plots L12, CH109C, CH138 of variety IRCA41 and the soils of plots L9, M9, CH158 of variety PB217 are acidic. On the other hand, the soils of plots M10, CH107B, CH42 of variety IRCA41 and the soils of plots CH135E, CH143, CH48 are neutral. Analysis of cation exchange capacity (CEC) showed that all plots had very high CEC, ranging from 30.3 to 44.9 30 Cmol/g. Also, on each plot, exchangeable base contents (K^+ , Mg^{2+} and Ca^{2+}) were high. Similarly, the organic fraction of the soils on all plots was high. The proportions of organic nitrogen were between 0.4 and 0.5%, and were very high. These soils are very rich in organic matter (7.3 to 8.7%). With the exception of the soils of plots CH109C, CH138, M10 and CH107B of the IRCA41 variety, which had high C/N ratios, all other soils had normal C/N ratios. However, assimilable phosphorus levels in soils from all plots ranged from 43 to 54.3 ppm. These soils are poor in assimilable phosphorus.

Table IV: Physicochemical properties of soils under CHC rubber cultivation

Plots	pH	C (%)	N (%)	C/N	MO (%)	CEC (Cmol/kg)	Mg2+ (Cmol/kg)	K+ (Cmol/kg)	Ca2++ (in %) (Cmol/kg)	Passs (ppm)
L12	6,3[a]	5,1[a]	0,5[a]	10,2[a]	8,7[a]	44,9[a]	2,3[a]	0,5[a]	3,9[c]	47,7[a]
M10	6,7[a]	4,8[a]	0,4[a]	12[a]	8,2a	32,2[c]	1,16[b]	0,2[a]	4,8[b]	54,3[a]
CH107B	6,7[a]	4,8[a]	0,4[a]	12[a]	8,2[a]	34,2[bc]	1,4[b]	0,2[a]	4,6[bc]	54,3[a]
CH109C	6,3[a]	5,1[a]	0,4[a]	12,75[a]	8,2[a]	31,2[c]	1,36[b]	0,2[a]	8,1[a]	54,3[a]
CH138	6,3[a]	4,8[a]	0,4[a]	12[a]	8,1[a]	37,5[b]	2,3[a]	0,4[a]	4,6[bc]	50[a]
CH42	6,7[a]	4,6[a]	0,5[a]	11,25[a]	8,7[a]	31,2[c]	1,16[b]	0,4[a]	4,6[bc]	48[a]
L9	6,1[a]	4,5[a]	0,4[a]	10,5[a]	8,2[a]	37,5[b]	2,5[a]	0,4[a]	8,4[a]	43[a]
M9	6,7[a]	4,2[a]	0,4[a]	11,5[a]	8,2[a]	37,3[b]	2,5[a]	0,4[a]	4,2[bc]	54,3[a]
CH135E	6,1[a]	4,6[a]	0,4[a]	10,5[a]	8[a]	32,5[c]	1,2[b]	0,4[a]	4,8[b]	51,3[a]
CH158	6,3[a]	4,2[a]	0,4[a]	10,5[a]	7,3[a]	36,5[b]	2,3[a]	0,4[a]	4,8[b]	45[a]
CH143	6,7[a]	4,2[a]	0,4[a]	10,5[a]	7,3[a]	30,5[c]	1,4[b]	0,4[a]	4,4[bc]	45[a]
CH48	6,7[a]	4,2[a]	0,4[a]	10,5[a]	7,4[a]	30,3[c]	2,5[a]	0,5[a]	4,9[b]	45,5[a]

C-carbon, N-nitrogen, MO-organic matter, Mg-magnesium, K-potassium, Ca-calcium, Pass-assimilable phosphorus

1.2. Sanitary condition of plantations visited

1.2.1. Symptoms observed on rubber plants

Three types of symptoms were observed. These were leaf desiccation, root rot and trunk splitting (Figure 8). These symptoms were observed in all plots. The average number of diseased plants in all plots ranged from 39 to 167 plants (Figure 9). For the IRCA41 variety, the plants showing the symptoms cited were 140; 77; 39 and 65 plants respectively according to plots in 1998, 2009, 2015 and 2017. The number of diseased rubber plants in IPB217 plots was 158 in 1998, 187 in 2009, 140 in 2015 and 125 in 2017. The PB217 variety had the highest number of diseased seedlings. The number of diseased seedlings decreased with the age of the plots.

The number of dead rubber trees on all plots (2 to 6 plants) was low compared with the sample size (510 plants) for each plot (Figure 10). The 1998 and 2009 plots had virtually the same number of dead plants, regardless of variety. The 2015 and 2017 plots of the IRCA41 variety had the highest number of dead plants.

Figure 8: Rubber plants showing symptoms

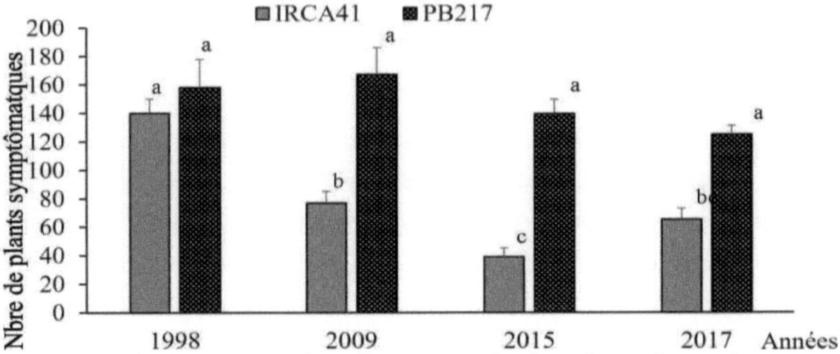

Figure 9: Average number of symptomatic plants by variety and age

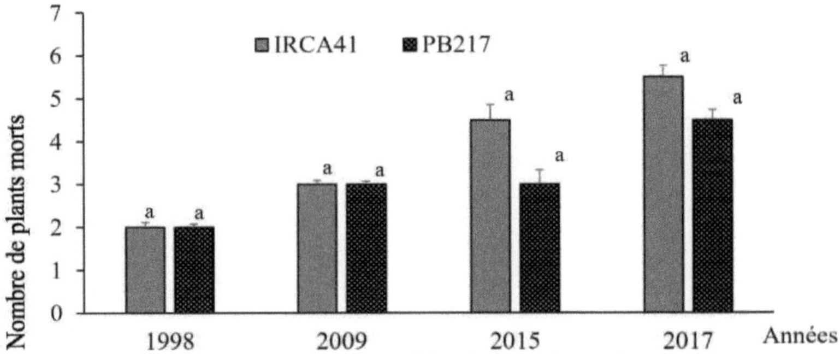

Figure 10: Average number of dead plants by variety and age

1.2.2. Microorganisms present in infected organs

Analyses of infected plant organs revealed four types of germ: Bacillus, enterobacteria, fungi and aeromesophilic germs (AMG) (Figure 11). Bacillus loads in organs of variety IRCA41 ranged from 5.103 to $2.7.10^4$ CFU/ml. Those of PB217 ranged from 2.103 to $1.8.10^4$ UFC/ml. Enterobacteria loads for IRCA41 ranged from 2.5.103 to 1.5.104 CFU/ml, while those for PB217 were between 4.103 and 2.3.104 CFU/ml. Fungal loads in the organs of variety IRCA41 ranged from 2.3,103 to 5.10^4 CFU/ml. In the PB217 variety, fungal loads in infected organs ranged from $1.3.10^4$ to 1.8.105 CFU/ml. GAM loads for IRCA41 ranged from $1.5.10^4$ to $7.5.10^4$ CFU/ml, while those for PB217 ranged from 2.10^{5} to $2.5.10^{5}$ CFU/ml. The leaf organs of this variety showed a higher *bacillus* load than those of the roots and barks. In terms of enterobacteria and fungi, the bark organs of the PB217 variety had the highest load, while only the GAM load was higher in the leaf, root and bark organs of the PB217 variety. Whatever the microorganism, infected root organs showed higher loads than leaves and stems for both varieties.

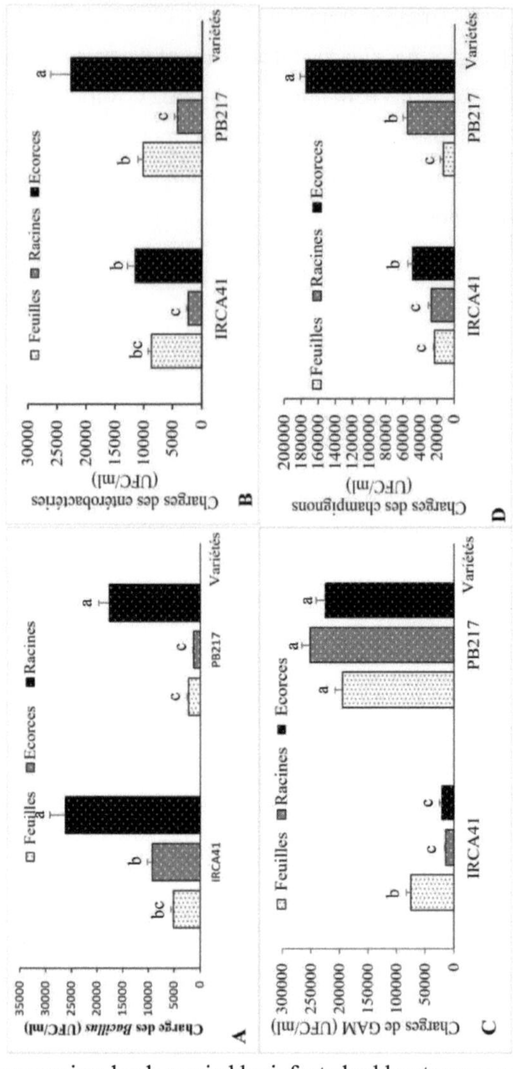

Figure 11: Microorganism loads carried by infected rubber tree organs, by variety

A- Bacillus; B-Enterobacteria; C-GAM; D-Fungi

1.3. Activities of arbuscular mycorrhizal fungi

1.3.1. Abundance of mycorrhizal fungi propagules in the soil

CMA spore densities in all plots ranged from 3.7 to 4.9 spores/g. For variety IRCA41, densities were 3.7; 3.8, 4.9 and 4 respectively for plots in 1998, 2009, 2015 and 2017. For the PB217 variety, densities were 4.7; 4; 3.9 and 4.8 spores/g respectively for the 1998, 2009, 2015 and 2017 plots (Figure 12).

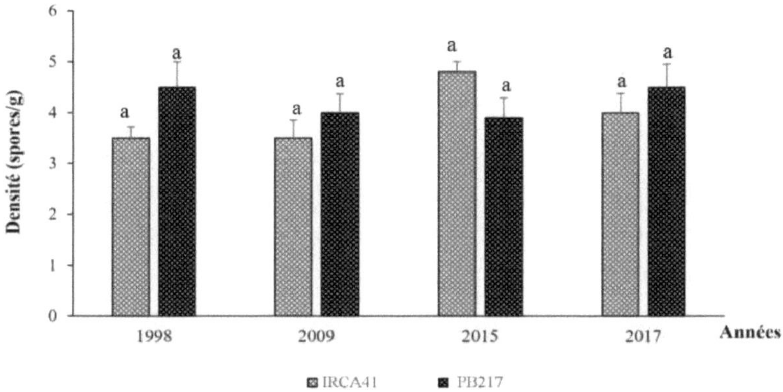

Figure 12: Density of mycorrhizal fungus spores in the soils of the Compagnie Hévéicole du Cavally rubber plots.

1.3.2. Mycorrhizal colonization as a function of year and variety

Mycorrhizal intensities in rubber roots from IRCA41 plots were 1.5% in 1998, 3.5% in 2009, 3.3% in 2015 and 3.3% in 2017. Those for roots on IPB217 plots were 2.5% in 1998, 3.7% in 2009, 3.9% in 2015 and 3.7% in 2017and 3. Whatever the variety and age, mycorrhizal density remained low (Figure 13).

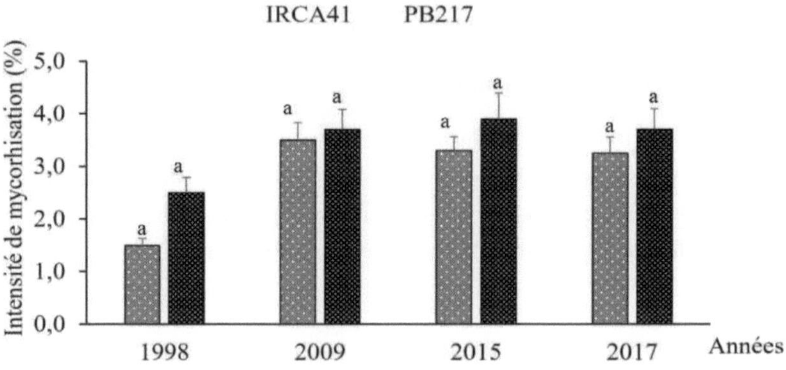

Figure 13: Mycorrhizal intensity in rubber roots as a function of age and variety

1.4. Plot production performance

Plots dating back to 1998, i.e. 22 years old, achieved 2020 yields of 2469 t/ha for variety IRCA41 and 2593 t/ha for variety PB217. The 2009 plots (11 years old) of variety IRCA41 have a yield of 1204.7 t/ha and those of variety PB217 have a yield of 1327.4 t/ha (Figure 14).

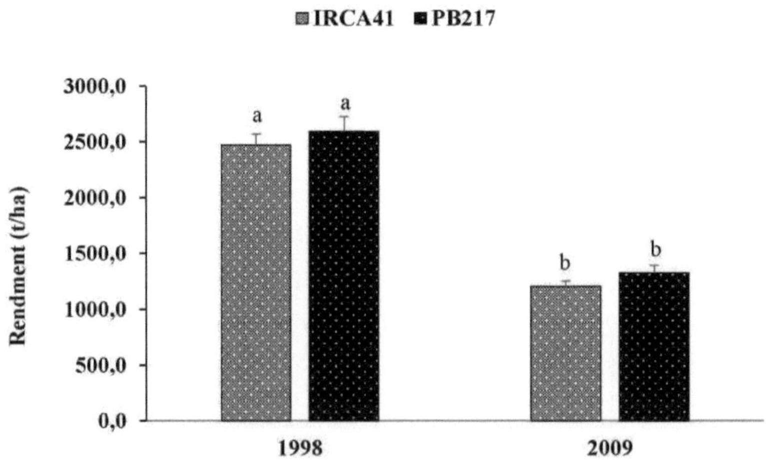

Figure 14: Comparison of 1998 and 2009 variety yields

1.5. Interaction between mycorrhization, rubber plant health and soil quality

Axes A1 and A2 express 64.72%, and are more than sufficient to characterize the relationship between physicochemical properties (OM, CEC, K^+ , Ca^2 +, Mg^2 +, PH_water and assimilable phosphorus), mycorrhizal intensity and the number of dead plants. Labeled individuals are those with the greatest contribution to the construction of the plan. Labeled variables are those best represented on the plan. This graph defines 4 classes (Figure 15). Class 1 is made up of individuals such as IRCA41_1998, PB217_2009 and PB217_1998. This group is characterized by: variables whose values do not differ significantly from the mean. Class 2 is made up of individuals such as IRCA41_2017. This group is characterized by: variables whose values do not differ significantly from the mean. Class 3 is made up of individuals such as PB217_2017. This group is characterized by variables whose values do not differ significantly from the mean. Class 4 is made up of individuals such as IRCA41_2015. This group is characterized by: variables whose values do not differ significantly from the mean. It can be seen that mycorrhization intensity has no influence on the other variables, notably plant morbidity and mortality (Figure 16).

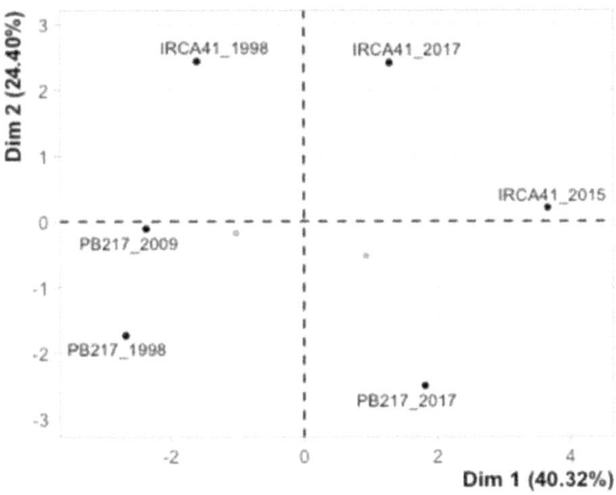

Figure 15: Hierarchical ascending classification (Plot of PCA individuals)

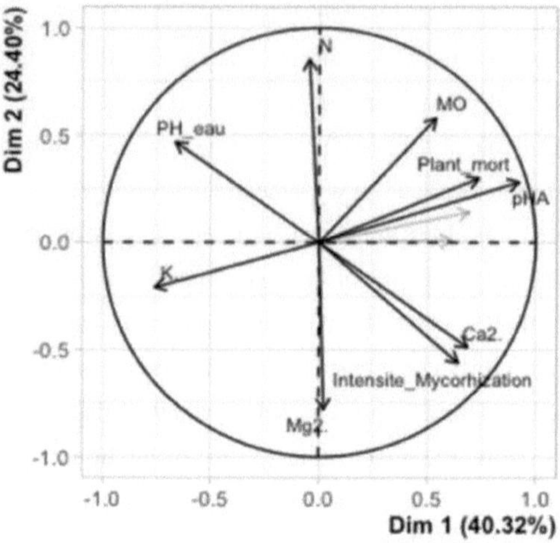

Figure 16: PCA of correlations between physicochemical properties, mycorrhization intensity and plant health (correlation circle in the FI X F2 factorial plane).

2. Discussion

The soil analysis carried out in this study showed that these soils have sandy-clay textures. This granulometric composition is more favorable to good structural stability for rubber tree cultivation. This assertion is endorsed by Obouayeba *et al.* (2006), who showed that for good rubber tree growth and production, a sandy-clay textured soil is required, deep and not waterlogged for at least one meter in depth, and flat or with a gentle slope (less than 8%) to avoid erosion.

The chemical parameters analyzed reveal that Cation Exchange Capacity values, organic matter contents and exchangeable bases are high, which explains the good chemical composition of soils under rubber cultivation. However, our results show that the mycorrhizal potential of the soils in these plots is low: mycorrhizal propagule density and mycorrhization intensity in the roots are low. The mycorrhizal potential of endogenous arbuscular mycorrhizal fungi would be inhibited by the use of chemical fertilizers. Indeed, these rubber-growing conditions would have a negative influence on the dynamics of arbuscular mycorrhizal fungi communities. In such cases, the role of mycorrhizal symbiosis is greatly reduced or even eliminated by mineral fertilization. Guerbault (2009) rightly asserted that increasing the concentration of mineral nitrogen and phosphorus in the soil solution increases protein synthesis and the synthesis of phosphorylated compounds by plants, leading to a decrease in soluble sugar content and mycorrhizal intensity in the roots. According to Kahiluoto *et al* (2009), in degraded tropical ecosystems, the adoption of conventional methods such as ploughing, mineral fertilizers or pesticides could lead to profound dysfunctions in the agroecosystem, likely to call into question its ability to satisfy the expected ecosystem services (provisioning, regulating, cultural and social services). For other authors, the application of pesticides to soils affects the mycorrhizogenic potential of vesicular and arbuscular mycorrhizae and the level of root infection. Gemma *et al* (2002) have attributed this reduction in the diversity and richness of vesicular and arbuscular mycorrhizal species to various agricultural practices that disrupt soil functioning.

Principal component analysis of the correlations showed that there was no correlation between the physicochemical properties, mycorrhizal intensity, yield and health status of rubber seedlings. This could also be explained by the reduced activity of arbuscular mycorrhizal fungi. In addition, there was no difference in production for any variety. This could also be explained by the lack of influence of arbuscular mycorrhizal fungi. Furthermore, the low rate of diseased or dead plants is more closely linked to good plant nutrition through the use of fertilizers and pesticides. Under the conditions of the Compagnie Hévéicole de Cavally (CHC) rubber plantations, the expression of Arbuscular Mycorrhizal Fungi is compromised.

CONCLUSION AND OUTLOOK

The present study focused on arbuscular mycorrhizal fungi, symbionts of crop roots. The aim was to determine the potential of endogenous CMAs under rubber cultivation in the agroecological soils of the Compagnie Hévéicole de Cavally (CHC). At the end of this work, it should be noted that the soil of the Compagnie Hévéicole de Cavally has a sandy-clay texture with a good chemical composition, characterized by the strong presence of cation exchange capacity, nitrogen, organic matter and exchangeable bases. Secondly, the mycorrhizal potential of endogenous CMAs is low in the roots of the IRCA41 and PB217 rubber clones, as a result of the company's cultivation practices. Finally, endogenous CMAs have no influence on the cultural performance and health status of rubber seedlings in this company's plots.

However, to ensure the sustainability of rubber cultivation in this region of Côte d'Ivoire, it would be worthwhile to reduce chemical inputs and progressively instruct biological fertility, in particular CMA. Given the ecological importance of arbuscular mycorrhizal fungi in ecosystems, it would be necessary to further characterize the structure of these microbial communities associated with each of the two varieties (IRCA41 and PB217) using molecular biology techniques, in order to gain a better understanding of how they function and thus enable inocula to be made from endogenous CMA strains.

REFERENCES

Angui K.T.P., Assié K.H., Djéké D.M., Tamia A.J. & Tié B.T. (2010). Morphological, physical and chemical characteristics of soils under different land use types. (Eds) Belowground biodiversity in Oumé (Mid-west Côte d'Ivoire). pp. 11-25.

Attobra A. (2010). APROMAC. Latest development of natural rubber plantations in Côte d'Ivoire. 15 p.

Bremner J.M. 1960. Determination of nitrogen in soil by the Kjeldahl method. *Journal of Agricultural Sciences,* 5 5: 1-3 3.

Brigitta S. & Sumalan R. (2011). The influence of arbuscular mycorrhizal fungi on ornamental characters of Tagetes patula L. Journal of Horticulture Forestry and Biotechnology, 15(1): 170- 174.

Burle (1961). Carotenoids. Spectroscopy. Birkhaüser. *Basel,41*: 7-23.

Calvel G. & Villemin P. 1986. *Interprétation des analyses de terre.* France, SCP A. 24p.

CNRA (2015). Center National de Recherche Agronomique. Online data from website as of February 16, 2015.

Compagnon P. (1986). Natural rubber. Biologie-culture-production. Maisonneuve et Larose Edition Paris. 595 p.

Dare M.O, Abaidoo R.., Fagbola O. & Asiedu R. (2012). Effects of arbuscular mycorrhizal inoculation and phosphorus application on yield and nutrient uptake of yam. *Communications in Soi! Science and Plant Analysis,* 41: 2729-2743.

Dare M.O., Abaidoo R.C., Fagbola O. & Asiedu R. (2013). Genetie variation and genotype x environment interaction in yams (Dioscorea spp.) for root colonization by arbuscular mycorrhiza. *Journal of Food, Agriculture and Environment,* 6: 227-233.

Demol J. (2002). Amélioration des plantes, application aux principales espèces cultivées en région tropicale. Presse agronomique de Gembloux, : 494-509.

Dos Santos, V.L., Muchoveg, R.M., Uchoveg, R.M., Borges, A.C., Neves, J.C.L. & Kasuya, M.C.M. (2001). Vesicular-arbuscular/ectomycorrhiza succession in seedlings of Eucalyptus spp. Brazilian Journal of Microbiology, 32: 81-86.

Duponnois, R., Ouahmane, L., Kane, A., Thioulouse, J., Hafidi, M., Boumezzough, A., Prin,

Y., Baudoin, E. & Dreyfus, B. (2011). Nurse shrubs increased the early growth of Cupressus seedlings by enhancing belowground mutualism and soil microbial activity. Soil Biology & Biochemistry, 43: 2160-2168.

Dusotoit-Coucaud A. (2012). Physiological and molecular characterizations of sugar and polyol transporters of laticifere cells in "Hevea brasiliensis", in relation with latex production. PhD thesis, Université Blaise Pascal, (Iles de France, France), 349 p.

Egerton-Warburton, L. & Allen, M. F. (2001). Endo- and ectomycorrhizas in Quercus agrifolia Nee. (Fagaceae): patterns of root colonization and effects on seedling growth. Mycorrhiza, 11 : 283290.

Elsen A., Gervacio D., Swennen R. & De-Waele D. (2008). AMF-induced biocontrol against plant parasitic nematodes in Musa sp a systemic effect. Mycorrhiza: Doi 10.1007/s00572008-0173-6.

FAOSTAT (2020). . *Faostat Agricultural database-Agricultural Production.* Retrieved January 28, 2020, from http://www.fao.org.

Friberg S. (2001). Distribution and diversity of arbuscular mycorrhizal fungi in traditional agriculture on the Niger inland delta, Mali, West Africa. CBM: Skriftserie, 3: 53-80.

Gadkar V. & Rillig M.C. (2006). The arbuscular mycorrhizal fungal protein glomalin is a putative homolog ofheat shock protein 60. Federation of European Microbiology societies Letters, 263: 93-101.

Gemma J.N., Koske R.E. & Habte M. (2002). Mycorrhizal Dependency of Some Endemic and Endangered Hawaiian Plant Species. *American Journal of Botany.* 89(2) : 337-345

Gianinazzi-Pearson V. & Smith S.E. (1988). Physiological interactions between symbionts in vesicular- arbuscualr mycorrhizal plants. *Annual Revue of Plant Biology,* 39: 221-244.

Guerbault, S. (2009). Mycorrhizae: tools for sustainable horticulture and agriculture. *Jardins de France,* 497: 19-25.

Huang S.P. & Cares J.E. (2004). Echantillonnage (Méthodes), Restitution du séminaire d'embussions, Kenya, Document 4, 7 p.

INVAM (2018). International culture collection of Vesicular Arbuscular Mycorrhizal fungi. http/www.invam.caf.wvu.edu. ; Downloaded January 01, 2021.

Kahiluoto H. & Vestberg M. (1998). The effect of arbuscular mycorrhiza on biomass production and phosphorus uptake from sparingly soluble sources by leek (Allium porrum L.) in Finnish field. Soil. Biology, Agriculture and Horticulture, 16: 65-85.

Kormanik P.P. & Mc Graw A.C. (1982). Quantification of vesicular arbuscular mycorrhizal in plant roots. American phytopatological society (Eds). *Methods and principles of mycorrhizal research.* Minnesota, USA. Pp. 37-45.

Lambers H, Mougel C., Jaillarrd B & Hinsinger P. (2009). Plant-microbe-soil interactions in the rhizosphere: an evolutionary perspective. *Plant Soil.* 321: 83-115.

Li H.Y., Yang G.D., Shu H.R., Yang Y.T., Ye B.X., Nishida I. & Zheng C.C. (2006). Colonization by the Arbuscular Mycorrhizal Fungus Glomus versiforme induces a Defense Response against the Root-Knot Nematode Meloidogyne incognita in the Grapevine (Vitis amurensis Rupr), which includes Transcriptional Activation of the Class III Chitinase Gene VCHJ. *Plant and Cell Physiology,* 4(7): 154- 163.

Liu J., Maldonado M., Lopez-Meyer M., Cheung F., Town C.D. & Harrison M.J. (2007). Arbuscular mycorrhizal symbiosis is accompanied by local and systemic alterations in gene expression and an increase in disease resistance in the shoots. *The Plant Journal,* 50: 529-544.

Lovelock C.E. & Ewel J.J. (2005). Links between tree species, symbiotic fungal diversity and ecosystem functioning in simplified tropical ecosystems. New Phytologisl, 167: 219228.

Lozupone C.A. & Klein D.A. (1994). Molecular and cultural assessment of chytrid and Spizellomyces populations in grassland soils. Mycologia, 94: 411-420.

MINA (2002). Ministry of Agriculture, Ivory Coast. Recensement National de I'
Agriculture, 22 p.

Morton J.B. & Benny G.L. (1990). Revised classification of arbuscular mycorrhizal fungi (Zygomycetes). A new order, Glomales, two new suborders, Glomineae and Gigasporineae, and two new families, Acaulosporaceae and Gigasporaceae, with an emendation of Glomaceae. Mycotaxon, 37: 471-491.

Muchane M.N., Muchane M., Mugoya C. & Masiga C.W. (2012). Effect of land use system on Arbuscular Mycorrhiza fungi in Maasai Mara ecosystem, Kenya. *African Journal of Microbiological Research,* 6(17): 3904-3916.

Ndong A.N. (2007). Importance of rubber clone association in improving rubber production on industrial plantations in Gabon. *Tropicultural,* 25(2) :66-69

Neobot (2020). Geolocation of zagné. Topographic map of Ivory Coast, 4 p.

Obouayeba S., Gnagne J.M., Wahounou P.J. & Boko C. (2006). Bien cultiver l'hévéa en Côte d'Ivoire CNRA, fiche technique de l'hévéa, 4 p.

Oehl F., Laczko E., Bogenrieder A., Stahr K., Bosch R., van der Heijden M. & Sieverding E. (2010). Soil type and land use intensity determine the composition of arbuscular mycorrhizal fungal communities. Soif Biology and Biochemistry, 42(5): 724-738.

Olsen S.R. (1952): Measurement of surface phosphorus on hydroxylapatite and phosphate rock with radiophosphorus. *The Journal of Physical Chemistry,* 56: 630-632.

Pansu M. & Gautheyrou J., (2003): Mineralogical, organic and mineral soil analysis. Montpellier, France: Springer-Verlag, 993 p.

Panthou E. (2013). Michelin plantations in Vietnam. Presse agronomique de Galipot, 125-210.

Phillips J.M. & Haymann D.S. (1970). lmproved procedure for clearing roots and staining parasitic and vesicular arbuscular mycorrhizal fungi for rapid assessment of infection. *Transactions of the British Myco/ogica/Society,* 69: 275-280.

Piéri C., Pierret, P. & Landais, E. (1998). L'analyse des trajectoires des exploitations agricoles. A method for updating typological models and studying the evolution of local agriculture. *Economie Rurale* 228: 35-47.

Redecker D., Hijri I. & Wiemken A. (2003). Molecular identification of arbuscular mycorrhizal fungi in roots: perspectives and problems. *Folia Geobotany,* 38: 113-124.

Redecker D. & Raab P. (2006). Phylogeny of the Glomeromycota (arbuscular mycorrhizal fungi): recent developments and new gene markers. Mycologia 98 : 885-895.

Rillig M.C. & Steinberg P.D. (2002). Glomalin production by an arbscular mycorrhizal fungus: a mechanism of habitat modification? *Soil and Biology Biochemistry,* 34: 1371-1374.

Selosse MC. & Le Tacon F. (1999). Symbiosis strategy, Science et avenir. *Indian Journal of hill farming,* Vol. 15, N°1, 15-17.

SIPH (2013). Internal control procedures and principles for determining compensation and benefits of any kind granted to corporate officers. Annual report, 199 p.

Schübler A., Schwartzzott D. & Walker C. (2001). A new phylum, the Glorneromycota: phylogeny and evolution. *Mycological Research*, 105: 1413-1421.

Smith S.E & Read. (1997). Interactions of nematodes with mycorrhizal fungi. In: Veech J.A. & Dickson D.W. (Eds). Yistas on nematology. Society of Nematologists, Hyatsville, Maryland, USA, pp. 292-300.

Smith S.E & Read D.J. (2008). Mycorrhizal symbiosis. Third Edition, 5, 769- 787.

Tchabi A., Coyne D., Hountondji F., Lawouin L., Wiemken A. & Oehl F. (2008). Arbuscular mycorrhizal fungal communities in sub-Saharan Savannas of Benin, West Africa, as affected by agricultural land use intensity and ecological zone. Mycorrhiza, 18: 181195.

Trouvelot A., Kough J.L & Gianinazzi-Pearson V. (1986). Measuring the VA mycorrhization rate of a root system. Search for estimation methods with functional significance. In: physiology and genetics aspects of mycorrhizae, 35(6): 217-221.

Voko B.R.D.R, Ahonzo-Niamkey L.S. & Zeze A. (2013). Impact of physicochemical properties of cassava-growing soils on the abundance and diversity of arbuscular mycorrhizal fungi communities in the Azaguié agroecological zone, Southeastern Côte d'Ivoire. *Agronomie Africaine* 25 (3): 251 - 264.

Voko B.R.D-R. (2014). Agroecology of endomycorrhizal fungal communities of three cassava (Manihot esculenta, Crantz) growing areas in Côte d'Ivoire. Doctoral thesis. UFR Biosciences. Université Félix Houphouët-Boigny, (Abidjan, Côte d'Ivoire). 164 p.

Vos C.M., Tesfahun A.N., Panis B., De Waele D. & Elsen A. (2012). Arbuscular mycorrhizal fungi induce systemic resistance in tomato against the sedentary nematode Meloidogyne incognita and the migratory nematode Pratylenchus penetrans. Applied Soil Ecology, 61: 1-6.

Walker C., Mize W. & McNabb H.S. (1982). Populations of endogonaceus fungi at two populations in central Iowa. *Canadian Journal of Botany,* 60: 2518-2529.

Walkley A. & Black I.A. (1934). An examination of the Degtjareff method for determining soil organic matter and a proposed modification of the chromic acid titration method. *Soil Sciences,* 37: 2938.

Wahounou P.J. (2010). Formation à la lutte contre Fomes en plantation d'hévéa Formation, FIRCA project. 15 p.

Yagoob H. (2015). Academic journals. Journal of Environmental science and technology full length research paper. The effects of arbuscular mycorrhizal fungi and phosphorus levels on dry matter production and root traits in cucumber (Cucumis sativus L.), 9(2): 65-70.

Yao K.M., Konaté S., Angui K.T.P., Tondoh E.J., Tano Y., Abbadie L., & Benest D. (2010). Effects of land use types on soil organic and carbon and nitrogen dynamics. Tondoh (Eds) Belowground biodiversity in Oumé (Mid-west Côte d'Ivoire). pp. 11-25.

Summary

Hevea *(Hevea brasiliensis), a* plant mainly grown for its rubber-rich latex, used as a raw material in various industries. The valorization of microorganisms present in the soil represents a sustainable alternative for improving soil fertility. The general objective was to determine the potential of endogenous arbuscular mycorrhizal fungi in soils under rubber cultivation in the agroecological soils of the CHC in western Côte d'Ivoire. Granulometric analysis showed that the Compagnie Hévéicole de Cavally soil has a sandy-clay texture, and chemical analysis revealed a good chemical composition characterized by cation exchange capacity ranging from 30.3 to 44.9 Cmol/g; nitrogen proportions ranging from 0.4 to 0.5% and organic matter from 7.3 to 8.7%; C/N ratios ranging from 10 to 12.75. Spore densities in all plots ranged from 3 to 4.5 spore/g, with mycorrhizal intensities ranging from 1.5 to 4. The continuous use of chemical fertilizers in rubber-growing soils greatly increases the content of organic matter and exchangeable bases, which in turn reduces the activity of arbuscular mycorrhizal fungi in the roots of the two rubber clones IRCA41 and PB217.

Keywords: Hevea; fertility; mycorrhizal; fertilizer; Ivory Coast

MIX
Papier aus verantwortungsvollen Quellen
Paper from responsible sources
FSC® C105338

Printed by Books on Demand GmbH, Norderstedt / Germany